Overcoming Math Anxiety

Becoming Successful in Math and Statistics

Custom Publishing

New York   Boston   San Francisco
London   Toronto   Sydney   Tokyo   Singapore   Madrid
Mexico City   Munich   Paris   Cape Town   Hong Kong   Montreal

Cover Art: Courtesy of PhotoDisc/Getty Images, Corbis

Printed in the United States of America

30 29 28 27 26 25

2008360893

MH

Printed in the United States of America

**Pearson
Custom Publishing**
is a division of

www.pearsonhighered.com

ISBN 10: 0-555-05117-X
ISBN 13: 978-0-555-05117-7

CONTENTS

PREFACE

Success Skills in Math and Statistics is a practical guide for students who want to succeed. This book not only provides specific tips and techniques to improve study habits, but also helps students understand how their beliefs and attitudes are related to their success in math and statistics.

In writing this book, the author wanted to address two fundamental questions: (1) Why do so many students struggle in math? and (2) How can we help them succeed? The following components of this book describe its usefulness to students:

- **Balanced Approach** ***Success Skills in Math and Statistics*** addresses not only *why* so many struggle with math, but also *how* we can become more successful in math.

- **Readability and Usability** We were determined to keep this book short and to the point so it could be easily used in a math or statistics course. This book is also substantive enough to be used as a primary text in a lab or study skills course.

- **Useful Time Management Content** An entire chapter is devoted to time management. So many times good students start off well only to realize they've over-scheduled themselves. This chapter provides practical tips and strategies including the School Rule and how to budget time.

- **"Try This!" Exercises** These exercises are designed to help students develop success skills like time management, note-taking, attendance and participation. They can be used as ice-breakers on the first day of class, warm-ups, homework, or diagnostics.

- **Successful Test Preparation Strategy** While there is no perfect approach that can guarantee perfect scores on every quiz and test, there are habits that successful students have used over time. Simple straightforward methods used by successful students are captured in this book.

CHAPTER 1:
INTRODUCTION

The Problem

Let's get straight to the point. On average half of the students taking developmental math classes (Prealgebra, Introductory Algebra, or Intermediate Algebra) or statistics do not pass on their first attempt. The other half don't enjoy the experience even if they did pass. Either way, student success is not occurring. Believe it or not, most instructors feel every student has the potential to successfully complete a college math and statistics course and yet so many students are missing their opportunity.

Can you identify with what these students are saying?

- I'm an A/B student in all my other classes. I try my best, but I just can't get math. I understand what the teacher says in class, I take good notes, I study hard, but when I get home and try the problems on my own, I don't understand how to even begin.

- I've never been good at math ever since I can remember. I was told by my third grade teacher that I had some kind of learning disability in math and that I should just learn to live with it. Ever since then, I've just surrendered to the reality that I'm not good at math, so I do what I can to avoid it.

- I'm a returning adult student and it's been longer than I care to admit since my last math class. I don't remember much of anything from my last math class and I'm afraid of looking stupid in front of other students and my professor.

I hear similar messages every semester. If you can identify with these remarks in any way, then you are not alone and this book can help.

Math and statistics are not like other subjects. They usually require nearly twice the time of most college courses. Math is cumulative, so we need to remember old concepts while learning new ones. Math requires daily practice. We need to make a commitment to attend and participate in every class. This is difficult for many students who don't like math or statistics. Let's face it, who wants to invest a lot time and effort into something we don't like? The good news is if we get it right the first time, then we don't need to repeat a course and we just might enjoy our success in the process.

The Solution

This book is written for you, the student. It is intended to be a practical guide for your success. You might feel anxious about your class and wonder what the outcome will be. The important thing to realize is that the solution to your success is within your reach. You can succeed in this class. You need to believe you can achieve!

The key to succeeding in math is having the right approach to studying and learning.

Observe that nothing is said here about how mathematically gifted we need to be. Success in math is more about knowing how to study math.

If a student approaches me and says "I'm just not any good at math," I ask to see their notebook. I'm often not surprised to see a confusing sequence of words, phrases, formulas, and a stack of loose papers shoved in the back of their notebook. While this approach may not hurt your success in other classes, it will affect your success in your math classes. Wondering why an poorly organized notebook doesn't help in math would be like wondering why a poorly assembled computer doesn't run well.

When a student comes to me and says "I keep trying, but I'm just not getting it," I ask them about their study habits. They explain how they can only study on the weekends. In math, that's a lot like expecting a professional football player to perform well on game day without practicing or warming-up during the week. Common sense would tell you this athlete would either get injured or make mistakes on the field (or both). In math, regular and consistent practice is a must or you will find yourself making mistakes and injuring your success. How you approach studying and learning math is much more important than how smart you mathematically gifted you may be. Not realizing this will make success more difficult to achieve.

This book is useless if you **only** read it.

You need to put the ideas in this book into action. Do the "Try This!" exercises. They are designed to help you build good habits. In order to get the most out of the exercises, you should be open and honest. Take a look at the following exercise to give you a sense of where this book is going and how it might help make you more successful in math.

Try This! Get to know this book. Read the table of contents and then browse the pages. As you browse through the pages, write down what draws your attention and include the page number. Describe in your own words what you found useful at a glance. The following prompts may help guide your thoughts:

1. The following chapters in the Table of Contents are of interest to me:

_____ .

2. I would like to start by taking a closer look at the following chapters:

_____ .

3. The following pages caught my attention: _____

_____ .

4. At a glance, the following ideas in this book appear useful to me:

_____

_____ .

Overview

Top Ten Math Myths

Math myths are the root of our problems in math. Sometime in our lives we started to believe we weren't going to become a successful math student. In this chapter, we take a look at the ten most common math myths. By discussing these myths, we can better understand them and get on the road to success.

Why is Math Different?

Few would argue that there's something different about math. Many of us know this intuitively, but we can't put our finger on exactly what it is. So we just say it's harder and surrender to the inevitable challenges. In this chapter, we identify what makes math different. We will also examine how these differences have tricked us into believing that math is harder than it actually is.

Learning Styles

Students have different learning styles. In this chapter we talk about different learning styles and specific success skills in math that will help your learning style become an asset.

Math Anxiety

This chapter talks about why math anxiety occurs in so many students. We describe some of the most common symptoms of math anxiety and offer specific tips on how to overcome it, so you can realize your true success!

Getting Started on the Road to Success

Knowing what we want to accomplish increases the chance that we will reach our goal. In this chapter, we discuss how to set goals and how achieve them.

Time Management

Success requires change. That change is most often in how students use their time. It could mean you spend more time in study groups and less time in social groups. It could mean any number of different things, but one thing is certain; time is your most valuable resource and how you spend it will have an impact on your success in this class.

During Class and Note Taking

Math is an intellectual exercise and you should feel confident and intelligent when sitting in class. In this chapter, we look at ways to help you become a more effective note taker and better student during class.

Studying

An entire book could be written on how to study. Students often share how they spend so much time studying only to realize they've failed the test. In this chapter, we offer a variety of techniques on how to study more effectively.

Test Preparation and Test Taking

You have probably heard that you should avoid cramming for a test. You may also have heard that you should get a good night's sleep and eat a healthy breakfast before a test. You probably also knew that homework is one of the most important things you can do in math. However, you may not have known that you have examples of every test question before the test. You also may not have known that you should bring hard candy to every test. Test preparation and test taking require skill just like any sport or performing art. We wouldn't expect an athlete or musician to perform at their best without a daily routine of practice. This chapter discusses how to become a test prep and test taking star!

Try This! Answer the following questions honestly.

1. How often do you study math or statistics? How long is each study session on average?

   _____

2. Do you prefer studying alone or in groups?

   _____

3. How do you begin and end a study session?

_____

4. How do you prepare for quizzes and tests?

_____

5. How could you improve your approaches to studying and learning math or statistics?

_____

CHAPTER 2:
TOP TEN MATH MYTHS

"No one can make you feel inferior without your consent."

—Eleanor Roosevelt

Math Myths are just that – *Myths!* Although not true, many students come to accept them as true. These myths create negative self-fulfilling prophecies that hinder our ability to succeed in math. Becoming aware of these myths can help us overcome them.

By confronting these myths, we begin to work through them. Throughout my years of teaching, I have heard students share their math horror stories. These stories typically went back as far as first or second grade and left a long-lasting impression. It's amazing how powerful these early childhood experiences in math become. Talking about these myths can help us move through them.

At the beginning of each semester, I list these top ten myths on the board. I ask my students to write down which myths they can identify with and why. By discussing these myths, students realize they are not alone and they begin feeling more comfortable in the class.

2.1 Ten Math Myths

Myth 1: "If you struggle in math, then you must not be smart." How often have you thought this? I encounter bright students everyday who simply aren't familiar with the "language" of mathematics. If you visit a foreign country where you do not speak the native language, does that mean you're not smart? Of course not! It simply means you do not know the native language of that particular country. It's the same thing in mathematics (or statistics). Math is a foreign language that perhaps you haven't fully mastered yet. I firmly believe anyone can succeed in math given the necessary resources.

Myth 2: We don't need math. One of the most common questions we hear in mathematics is

"Why do we need this?"

If you don't use it, you lose it! Think about which arm you use more. If you are right-handed, you tend to use your right arm more (if you are left-handed, you tend to favor your left). Which arm is stronger? Learning math is not just about learning how to solve a particular exercise on a particular page in a particular book (although that is an important part of the process).

Learning math helps us to develop critical thinking skills, analytical skills, and to become better problem-solvers in every day situations. Whenever you break large overwhelming tasks into smaller more manageable ones, you are using problem-solving skills developed from mathematics. Thank your statistics instructors for the ability to read, interpret, and analyze information (or data) in newspaper articles and magazines. Math and statistics are used to interpret and analyze global issues in our modern society. People who demonstrate critical thinking and problem-solving skills have job security. We will always need people with these skills in math and statistics.

Myth 3: Math is boring. The reality is that math can be found in just about anything: art, music, science, politics, criminal investigations, current events, toys, cars, sports, and the list goes on! If you're interested in something, look for the math in it! I promise you; it's there!

Myth 4: Math is supposed to be hard. The other half of this statement could then follow "and I can't make it any easier!" Being organized can help you overcome this myth. Getting all your course materials ahead of time, knowing how to access tutoring services, reading the text book before class, and getting the contact information of at least two other students in your class are specific tips to help make math a little less difficult to master. Don't be afraid to ask for help. Your education is a major investment and you owe it to yourself to create every opportunity for your success.

Myth 5: Men are better than women at math. Actually, recent studies show the math achievement gap between men and women actually favors women (Birenbaum & Nasser, 2006). There are many brilliant female mathematicians who have made significant contributions to the field.

Myth 6: Math is not practical. According to Webster's Dictionary, practical means "adapted or designed for actual use." This goes back to one of the most common questions I hear from my students.

"When will I ever use this?"

The key to breaking down this myth is to *use* the math we learn. The problem is that we are not always shown *how* to use it. This can be very frustrating and I too share your frustration. Some text books are better than others and some instructors are better than others at showing how the math we are learning can be used. Learning how to use math may sometimes be about helping us to prepare for the next chapter. Sometimes we will use the math in a later course. There may be times when the math itself is not be used, but the critical thinking and problem-solving portion of our brain is benefitting. Math "is" practical, but it helps to know "how" we're going to use this so we're ready to use it when the time comes.

Myth 7: There is only one way to solve a math problem. Okay in some cases, this may be true. For example, if we're asked to add 1 + 1 = 2, there aren't too many different ways we can "solve" it. This myth often gives students "writer's block" on homework or tests. You think there's only one way to solve this problem correctly and you aren't sure where to start. Text books often try to give a general outline on how to problem solve, usually based on George Polya's *How to Solve It* (1957). There are as many different variations as there are text books. The point of bringing this up is to be comfortable in knowing that there will often be more than one way to solve a problem and that's okay!

Myth 8: Math is all about logic and numbers. I am a mathematician. The truth of the matter is I am not that good with numbers! In fact, a number of capable mathematicians share this fate for numbers. Of course math involves both logic and numbers, but there's much more to it than that! The next myth sheds a little light on what I mean here.

Myth 9: Math is not creative. There is a strong relationship between mathematics, art, and music. The Pulitzer Prize-winning book *Gödel, Escher, Bach: an Eternal Golden Braid*, discusses this relationship in great detail. In many instances, mathematicians are hired more for their creative problem-solving abilities and less for their ability to solve advanced theoretical math problems. Companies want people with some math in their college studies because it demonstrates a certain level of discipline and problem-solving skills. One of the most valued qualities in math *is* that it is creative.

Myth 10: Success in math can't be learned. Many students believe you either have it or you don't. The truth is the vast majority of math is learned through 90% perspiration and 10% inspiration. The three most effective ways to become successful in math or statistics are to practice, practice, and practice. Math is a skills-based subject, which means it must be practiced and not just read to be mastered. We will talk about this in more detail in Chapter 3: Why is Math Different?

These myths have negatively impact students' attitudes and habits in math over many years, resulting in math anxiety and lower grades. In Chapter 5: Math Anxiety, we discuss the anxiety resulting from these myths and how to help you overcome them.

Try This! Describe in your own words what each of the math myths mean to you. Indicate which of the myths you can identify with.

_____

_____

_____

_____

_____

CHAPTER 3:
WHY IS MATH DIFFERENT?

"Challenges are what make life interesting; overcoming them is what makes life meaningful."

—Joshua J. Marine

3.1 Why is math different?

How often have you asked yourself this question? You think of yourself as a good student and yet math seems to give you more trouble than any other subject. I've heard the following statement often from my students:

I can get A's in other classes, why not math? Why is math so hard for me?

There is a reason math seems harder for so many students.

Math is a cumulative skills-based subject. In other words, math is best learned through repeated practice and not by just reading your text or notes. To illustrate this point, consider sports, which are also skills-based. You need to practice to improve your abilities in any sport. Math is similar. You need to practice to improve your abilities in any math class. Can you relate to the following statement from one of my students?

I understand everything in class. But once I get home and try to do the homework, I don't understand anything!

This is fairly common. I try to remind my students that math is a skills-based subject. You won't really "get it" by just listening to class lectures, taking notes, and reading your text book. You need to *do* the math to learn it.

Knowing math is skills-based is only helpful if you act on it. Don't fall into the trap of thinking you understand something until you have done it on your own. Successful students study math every day. Avoid intensive study sessions once or twice a week. A lot of time can be lost in trying to remember what was discussed in a lecture several days ago. Practicing math every day in shorter study sessions is better than trying to do it all in one long intensive session each week.

Try This! Why do you think math or statistics is so challenging for you?

_____

_____

_____

_____

_____

CHAPTER 4:
LEARNING STYLES

"If everyone is learning alike then somebody is not learning."

—George S. Patton

We all learn in different ways. Generally speaking there are three types of learning styles: visual, auditory, and kinesthetic. Many students have strengths in more than one learning style. This chapter will help you determine which of the three learning styles is your strongest.

The majority of math or statistics math classes are taught using chalk boards, power points, and notes, all of which benefit the visual learner, but not the auditory or kinesthetic learner. While we can't expect our instructor to change the way they teach, we can adjust the way we study to fit our primary learning styles. Let's take a look at the following questions to figure out which learning styles you prefer. Then browse through the *success tips* to find what works best for you.

4.1 Visual Learning Style

Are you a visual learner?

1. Are visual aids, such as graphs, figures, and boxes in your text book particularly helpful?
2. Do you find it difficult to understand concepts when your instructor is talking, if they don't write them on the board?
3. When problem-solving, do you find it particularly useful to start by constructing a diagram or figure?
4. Do you like to highlight your text book or class notes?
5. Do you prefer taking notes to listening in class?

If you answered yes to a majority of these questions, then you are a visual learner. Visual learners are the most common of the three types of learners. Here are some tips for the visual learner.

Success tips for visual learners:

Tip #1: Rewrite your notes as soon as possible after your class lecture; this will help you remember important concepts and reinforce understanding.

Tip #2: Your notes are very important for you. Take extra time to make sure they're organized and complete (see Chapter 8). High lighting your notes and text are helpful during study sessions. Use a variety of different colors to help you categorize your notes.

Tip #3: Construct figures and diagrams when problem-solving.

Tip #4: Note cards are useful study tools.

Tip #5: Writing notes in your text or note book while you're reading your text will be useful.

4.2 Auditory Learning Style

Are you an auditory learner?

1. Do you prefer listening in class to writing notes?
2. Do you find written solutions difficult to follow unless someone is verbally explaining them?
3. Do you find yourself reading or writing aloud when you study?
4. Do you prefer discussing problems in study groups or with your instructor rather than studying alone?
5. Do you tend to remember what you hear better than what you see?

If you answered yes to most of these questions, then you are an auditory learner. Auditory learners are the second most common of the three learners. The following tips may be useful for auditory learners:

Success tips for auditory learners:

Tip #1: Bring a recording device to class. Ask your instructor if you can record the lecture while you're taking notes. (The majority of instructors will approve). Listen to your recorded lectures as you review your notes.

Tip #2: In addition, lecture CD-ROMs are often available through the textbook publisher at little or no additional charge. They are particularly useful for auditory learners since you can listen to the lecture repeatedly as you follow your notes or the text.

Tip #3: Get into a study group. The more you can discuss math with your fellow classmates, the better. If you're in an online class, get the phone numbers of several classmates and set up study sessions on the phone or via webcams.

Tip #4: When you are studying alone, read aloud (make sure you're studying where you can speak out loud without disrupting others).

Tip #5: Try using a shorthand system so you aren't writing more than necessary. For auditory learners in particular, the less time you spend writing, the more you listen, the more you retain what is being taught.

4.3 Kinesthetic Learning Style

Are you a kinesthetic learner?

1. Do you find it particularly useful to re-write your class notes or to take notes while you are reading your text book?
2. Do you need to frequently get up and walk around while you're studying math?
3. Do you enjoy giving presentations or speaking in front of a group?
4. Do you tend to doodle during meetings or in class?
5. Do you take frequent small breaks during study sessions?

If you answered yes to a majority of these questions, then you a kinesthetic learner. Kinesthetic learners are the least common of the three types of learners. The following tips may be useful for kinesthetic learners:

Success tips for kinesthetic learners:

Tip #1: As a kinesthetic learner you need to be active. Re-write your class notes as soon as possible after class.

Tip #2: You learn best by doing, not just reviewing. Taking good notes is the first step, but practicing problems repeatedly will help immensely.

Tip #3: You need to get up and move during your study sessions. Give yourself permission to take short breaks regularly. You may find it helpful to stand or walk during part of your study session. You may even want to sprint up and down stairs for a minute to help you get back in the mode of studying math.

Tip #4: A good study tool for kinesthetic learners is a Concept Map (see Chapter 9 for more information on how to build a Concept Map).

Tip #5: Study outside where you can move freely. Read your text or class notes on a tread mill or exercise bike. Bring your math work to a place where you can be in motion. A quiet library may not be the best place for you to study if you're a kinesthetic learner.

Try This! Indicate your primary and secondary learning styles, based on what you have read. Explain why you chose these two learning styles.

_____

_____

_____

_____

_____

Try This! Choose seven tips you feel would be most useful to you. Explain your reasoning.

_____

_____

_____

_____

_____

_____

_____

CHAPTER 5:
MATH ANXIETY

"Nothing can stop the person with the right mental attitude from achieving their goal; nothing on earth can help the person with the wrong mental attitude."

—Thomas Jefferson

I struggled in math a lot! I can remember one day being asked to figure out an answer on the board in front of the entire class. I froze! I knew the math, but I was too afraid of making a mistake. I didn't think I was smart enough. I wasn't asked to the board again. My confidence in math at that moment was shattered. The teacher told my parents I would need to repeat this class in the summer. I was devastated!

I'm sharing this because it illustrates how many of us (even your professors) have bad math-related memories. These negative memories affect our confidence and attitudes towards math, which may result in math anxiety. Getting the right mental attitude is crucial to overcoming math anxiety.

5.1 What is Math Anxiety?

"... feelings of tension and anxiety that interfere with the manipulation of numbers and the solving of problems in a wide variety of ordinary life and academic situations."

—Sheila Tobias

Math anxiety affects students of all abilities in a variety of ways. For some, there exists a fear of appearing "stupid," for others a fear of appearing too smart. Math anxiety is well documented and is one of the most significant obstacles to student success in a math or statistics course. Some common symptoms of math anxiety are discussed in this chapter.

5.2 Symptoms of Math Anxiety

Students who struggle with math anxiety often find themselves going to great lengths to avoid math. Others freeze-up under test conditions even though they are well prepared.

Avoidance
- Waiting to register for math classes until the latest possible semester
- Skipping class
- Not reading the text
- Cramming
- Procrastinating

Poor Performance
- Freezing up on in-class graded work

Most people who suffer from math anxiety can recall negative experiences sometime early in their math studies. Students have shared with me that these experiences were incredibly embarrassing, often resulting in their feeling insecure or even ashamed of their level of math abilities. To make matters worse, our culture perpetuates math anxiety by suggesting math is harder than most subjects. Math is not harder, but it is different (remember Chapter 3).

Prescription for Math Anxiety

Before going any further, review the Top Ten Math Myths (Chapter 2). Remember which myths you can identify with and jot them down on a piece of paper before you continue reading this section. I want you to begin facing your anxiety and really work on letting it go.

- **Practice math every day.** It is better to have a consistent routine of 20–30 minutes of math practice every day then to cram all of your study time into one long session. This can't be said too often. You want to make sure you set a realistic schedule to do your homework, study your notes, and read your text book. The more you practice, the better you will get.

- **Study efficiently.** Use the strategies in this book to help guide your study habits. It's not just the quantity of study time, but also the quality of the study time. Use your class notes, read your text book, prepare flash cards, practice your homework questions, study in small groups and individually. Try a variety of techniques to see which one(s) fit your learning style best.

- **Attend and Participate in Class.** Do not miss class. If for some reason, there is no way to avoid missing class be sure to contact the two students whose email addresses you got the first day or so of class. When you attend class, be sure to participate in class. Math is *not* a spectator sport! Participate in class; ask questions. Remember to read your text book before each class, so you can organize your questions in advance. Most instructors appreciate well-prepared "instructive" questions. If yours doesn't, maybe you should change instructors.

- **Get Organized.** While some students may be able to function in chaos, all will benefit from being organized. Since math and statistics are both so cumulative in nature, it is important to be organized so you can retrieve what you need. In Chapter 7, we will discuss *how* to become better organized. This is a life skill that will serve you well beyond math and statistics.

- **Practice Self-Assessments.** Use the examples and odd-numbered exercises (that have complete solutions) to write practice quizzes and exams. Use this technique regularly and in timed situations similar to your quiz and exam environment. It is usually better to use practice tests shortly after you have studied a particular section or chapter (not before). In Chapter 10, there is a section on how to use practice quizzes or tests to assimilate the real test.

"To accomplish great things, we must ... not only plan, but also believe."

—Anatole France

- **Believe you can Achieve.** Beliefs are powerful, words are powerful. Think about how powerful your negative beliefs are... Make them positive!). Your negative self-talk is more than mere self-talk. Students who are anxious about math or statistics often say negative things about their abilities in these courses.

  - I'll never be smart enough to get this math.
  - I'm just not good with numbers.
  - I don't need to know it; I just want to pass it.

  With negative self-talk, there is actually a chemical release in the brain that literally makes it more challenging for you to make connections between the parts of the brain that understand and carry out math problems.

  Math anxiety contributes to negative beliefs towards math. These negative beliefs and attitudes can result in significant obstacles to your academic success. Believing you can reach a certain goal is the first step towards achieving that goal (otherwise, why even try?) How can you do this? Begin right now! Regardless of any negative past experiences you have had in math, make the following **positive** commitments:

  - I can achieve success in this course.
  - I am smart enough to succeed in this course.
  - I am responsible for my success in this course.
  - I will invest the time and effort to succeed in this course.

- **Learn from your mistakes.** It is important to be patient with yourself. Know that even the best mathematicians make mistakes. In fact, the most successful people often say they learn more from their mistakes than from their successes.

"Failure is only the opportunity to more intelligently begin again."

—Henry Ford

Don't spite your mistakes; use them to become a more successful math student. Understand where things went wrong. Was it a careless mistake? Did you not understand the concept? Was the question confusing? If your mistake was careless and you understand why you made the mistake, then remember to slow down next time. Careless errors are often made when we rush. If you do not understand a concept, then review the concept in your text and in your notes. If you are still struggling with understanding the concept, then either ask a fellow classmate (in your study group) or ask your instructor. Remember to be as specific as you can when asking your instructor for help. The better your question is phrased, the better the answer will likely be.

- **Use your course materials and resources.** You will likely have a variety of course materials and resources at your disposal. Be sure to get them as early as you can. You want to become familiar with them before class if at all possible, so you know how to use them before the pace of your semester picks up. Some of the course materials and resources may include your text, student solutions manual, CD-ROMs or DVDs, study guides, tutoring, just to name a few. You are investing a lot of time, effort, and money into this course. You deserve full use and access to *all* of the materials and resources available for your course. Make the commitment to become familiar with these resources and determine which are best for your success.

Try This! Write a math and statistics timeline. Include the grades for each math course along your time line from your earliest to most recent math and statistics classes. Then use the timeline to write your math autobiography from your earliest experiences in math classes to your most recent experience. Include as much detail about your attitudes and beliefs as you can.

_____

_____

_____

_____

_____

_____

CHAPTER 6:
GETTING STARTED ON
THE ROAD TO SUCCESS

"More powerful than the will to succeed is the courage to begin."

—Unknown

Ease into school. If you're relatively new to higher education or if you have numerous life commitments, ease into school. By registering for this course, you have already demonstrated the courage to begin. It is also important to demonstrate the good judgment to succeed.

Be realistic and set an honest schedule (see Chapter 7: Time Management). Your college education is a significant investment. It is a marathon, not a sprint. The following tips are particularly helpful for returning adult students, but they can easily be applied by any student who lives a busy life.

- **Weekly planning.** The overwhelming majority of adult students say that time is their biggest problem. They don't have enough time in the week to accomplish everything they need to. Create weekly plans for the semester. Write down your known commitments and leave plenty of space for surprises. In Chapter 7, we discuss the importance of time management in more detail. Once you've created your weekly schedule for the semester, hang it up on the refrigerator or cork board where others who live with you will be able to see it. Include empty spaces in your schedule so others will know when you're available and for how long. Remember, your schedule also affects those in your immediate circle and the more they can understand your schedule, the more they can appreciate and support you through the semester.

- **Connect with other students in your class.** Reach out. Introduce yourself to other students in your class. There is an important bond established just being in the same class. Share email addresses or phone numbers with other students in your class. This will be helpful if you miss a class and need to obtain a copy of class notes; it is also a good way to build a study group. You may find some of your classmates to be long-term friends as you move through other classes. Creating a learning community of classmates and friends will improve your overall college experience.

- **Communicate with your instructor.** Believe it or not, your instructors are human beings too! They're likely managing a busy life – just like you. Reach out to your instructor and let them know who you are and what your interests are in college. If you approach them periodically for short visits or email them occasionally for minor questions, you'll find it easier to approach them for the more important and more difficult questions later on.

- **Communicate your academic goals with your employer.** Your college experience makes you a more valuable employee. Communicate your educational plans with your employer. Discuss both in general terms and specifically how your college experience will help you become a better contributor to the work environment. When given the opportunity express gratitude to your supervisor for their support. Your employer can become an important asset.

- **Synergize your life (school & work)** *Prior Learning.* Look for ways to bring your work and your education together. Some schools will give you college credit for your real-life experience. Your employer may cover your tuition (reimbursement or up-front). You may also receive time off to attend class. Try to include your work experience into your college experience. For example, if you are asked to complete a research project, you may try to include topics related to what you do at work.

6.1 Create a set of goals for the semester.

Many of us have idealized goals such as "I want to be successful," "I want to be financially secure," or "I want to be a good person." These are good goals to have, but they are also very general and difficult to know for certain whether they've actually been achieved. It is important to set goals for yourself that are specific, attainable, and cover different time frames (e.g., short-term, mid-term, and long-term goals).

Short-term goals are the ones you can achieve within a year. They are specific and measurable. For instance completing that challenging math requirement, spending a weekend volunteering at an elderly care home, or reconnecting with an old friend.

Mid-Term goals are those goals that can be accomplished in one to five years. These goals can include earning a college degree, paying off credit card debt, starting a new career. Mid-term goals should support your long-term goals.

Long-term goals are significant achievements that take anywhere from five years to an entire lifetime to achieve. These goals can include family related goals, financial goals, education or career goals. Some long-term goals may change, while others may remain constant. It is important to re-evaluate your long-term goals as they relate to your mid-term goals once in a while to make sure you're on the right track.

Writing down your goals improves your chances of achieving them. It also helps you see if you're on the right track or if you've starting moving away from your goals.

Remember to set attainable goals. You want to build on success. Your short-term goals should guide what you do and how you move through the semester. For instance, you could set a goal to get all your required materials before classes begin. You could commit to reading your text book before each class. You could set a goal to get at least two email addresses from students in your class. You could set a goal of completing your homework at least 1 day before class.

Try This! Write down five goals for your class. Be sure to set challenging yet realistic goals. These goals should be attainable so you can reward yourself for each success.

_____

_____

_____

6.2 Required and Recommended Course Materials

You should make every effort to obtain your required and recommended course materials before class begins. Getting your course materials in advance helps you get better prepared for your course. You can also get a sense of whether or not this will be a relatively easy, medium, or hard course so you can use the School Rule to figure out how much time you'll need to succeed in this course.

Most text books are available used at a reduced cost. As long as the book is functional and not marked up or highlighted, it's okay used. You don't want to rely on someone else's highlights or notes (you don't know how they did in the class). With the rising cost of required text books, it's becoming increasingly financially demanding on students and the decision of whether or not to invest in the "recommended" course materials can be an important one. I tend to err on the side of getting all the recommended materials; however, if you are really in a pinch and want to know whether or not the recommended materials really are important wait until the beginning of class and ask your instructor.

Typical required/recommended course materials that do not include a text book are:

- Three-ring notebook
- Graphing calculator (if permitted)
- Pencils
- Ruler, protractor, pencils
- Weekly planner

The three-ring notebook is a very important tool for success in your math or statistics class. In Chapter 8, we include a discussion on how to organize your note book. Some instructors may allow graphing calculators. Find out as soon as possible whether or not your instructor allows the use of a graphing calculator (or some other technology). If your instructor allows the use of a graphing calculator and you do not have one, get one. They can be useful checking devices and time savers. User manuals can be helpful, but tedious to read. The following link is a free interactive graphing calculator tutorial (provided by Pearson) http://prenhall.com/divisions/csm/app/calcv2/ (Note: TI-84 users can follow the TI-83 help).

Your instructor will give you a list of tools you will need for the course. She will not likely include a weekly planner on that list. A weekly planner will become your best friend when it comes to managing your time (see Chapter 7).

6.3 Your Community of Support (Student Services)

Ultimately you are responsible for your own success. As part of that responsibility, you should know how to access all the student services available through your institution. Some of these student services may include:

- Academic advising
- Childcare
- Computer labs
- Counseling centers
- Financial aid office
- Job placement office
- School catalog
- Student organizations
- Tutoring

This list is by no means complete, but it gives you a start. Remember that you are likely paying for these services through tuition and fees, whether you use them or not, so why not at least take a look!

In addition to institutional support, you need to create a community of support, which may include family, friends, colleagues, or even new friends. When you begin your college experience, you may find yourself in a situation where difficult decisions need to be made regarding who will be in your immediate community of support. Evaluate who offers you something towards your college success and who might cost you in the long run. This may mean losing some old friends for a while, and creating space for new friends.

6.4 Make a commitment to attend *and* participate in every class.

If this sounds like simple common sense, that's because it is! Successful students do not miss class. On the rare occasion they miss class, they immediately contact their classmates to find out what they missed. Make a commitment to attend every class. Also, make the commitment to participate in class. To help you participate more in class, sit in a *premier seat* (or at least a *good seat*) in your classroom (Chapter 8).

"The people who get on in this world are the people who... look for the circumstances they want, and, if they can't find them, make them."

—George Bernard Shaw

Remember that you are in control of your commitment to this course. Create the circumstances you need to succeed. Participation does not mean you need to have all the answers; it can mean that you have questions. Don't fall in to the trap of getting off to a good start, then skipping a class as a reward for your hard work. Once you miss a class, it becomes easier to skip additional classes and you begin falling behind. Making a commitment to attend every class is a hard habit to make; missing class is a hard habit to break.

6.5 Get help before you need it.

Know where the Math Tutor Center is (or how to access it if you're in an online class). Free tutoring services are often available through the university or through the publisher of the text you are using for the course. Try these free services before you even need help; they're free!

6.6 Read your text book before class.

Reading a text book is not like reading another book, magazine, or newspaper. You will read your text at a much slower pace. You will stop and start more frequently. You will need to take notes and write questions down if you do not understand something. In mathematics texts, examples are good resources for helping you with end of section exercises. It's a good idea to try solving the examples on your own before looking at the solution in the text. By reading your text before class, you can get a sense of the layout and design of your text. You can get familiar with the different features in the text (table of contents, index, bolded vocabulary terms, examples, exercises, and so on. The more you can read before class, the better prepared you will be to participate and to ask questions. Your professor will appreciate the time you invested and you will feel more connected to the lecture.

6.7 Ask questions.

Ask questions! The worst question is the one never asked. Don't be afraid to ask a question; chances are a number of your classmates have the same question and are afraid to ask it. Some instructors may not answer your question the way you wanted them to. One strategy is to come to class prepared (read your text book before class). If you say to your instructor "I don't understand the chapter..." you probably won't get the help you are looking for. If you say to your instructor "I was reading the section on solving quadratic equations and got a little confused on completing the square. Could you explain again how to use completing the square to solve quadratic equations and then give an example to the class?"

The better your question is, the better the answer will be.

CHAPTER 7:
TIME MANAGEMENT

"Success is the sum of small efforts, repeated day in and day out."

—Robert Collier

Successful students develop successful habits. Our time is our most precious resource. Once we spend it, we can not get it back. Often times we do not realize how much time school demands of us. We know how exhausted we feel. We understand that we're overextended in many directions. We feel tired all the time, but we don't really appreciate exactly how much time is required to succeed in school.

7.1 Student Motivation

"Winning isn't everything, but wanting to win is."

—Vince Lombardi

In other words, succeeding in your math class isn't as important as *wanting* to succeed. This may seem a little strange at first, but allow me a moment to explain. The focus here is on what is motivating you. The success of completing your math class comes after the class is over; however, your motivation to succeed is what will carry you through the entire semester. Motivation is what will guide your discipline to achieve your academic goals. If you are focused on the end of the course, you may find many missed opportunities throughout the semester. The bottom line is, only you can motivate yourself to do what is necessary to succeed. Reading this book or any other book on how to succeed won't help unless you are willing to put it into action. This chapter focuses on time management. An important first step towards getting a handle on time management is what I like to call the "School Rule."

7.2 The "School Rule"

Time management is arguably the most important factor in becoming a successful college student. Many students become overwhelmed with the social and academic pressures that come with being newly independent. Other students are faced with financial pressures to pay for college or support a family while they work through college. As a result of all the additional responsibilities and opportunities, students find themselves with little time left for study.

Regardless of your situation, you want to get a handle on where your time is being spent (or invested). It's not unlike balancing your check book. You wouldn't want to continue spending your money without logging your spending in the check book. If you don't log your spending, then you begin to overspend and checks bounce. If you don't begin to manage your time in college, you will overspend it and find your performance will suffer as a result. So how do you begin to budget your time? The following is a good rule to follow in school, called the "School Rule." This may be common sense to some and difficult for others.

The School Rule

To succeed in school, you should budget *two* to *four* hours outside of class for every hour spent in class.

Let's take a look at why most institutions consider 12 credit hours full-time. If you are registered for 12 credit hours in a semester, then you will be spending roughly 12 hours per week in *class*. Using the School Rule, you should budget a **minimum** of 36 hours per week total:

$$12 \text{ hours } in \text{ } class + \text{ hours } studying = 36 \text{ hours per week } total$$

This is the same time commitment as a full-time job. It is important to note here that the School Rule states that this is the minimum amount of time you should budget in order to succeed in school and not simply get by in school. In some cases, succeeding may mean you successfully completed a very challenging course, while in other cases, it may mean you have discovered something you love to study and have decided to pursue it further. Success is a relative term and should not be defined purely on a letter grade earned in a course. Another important thing to note here is the School suggests a range (2-4 hours). While some courses may only require 2 hours per hour in class, others may require up to four hours studying for every hour in class. In a typical 12 credit semester, more challenging courses could demand an additional 10–20 hours per week on top of the 36 hours already budgeted.

Try This! Create your weekly schedule for the entire semester. Use the *School Rule* to estimate the number of hours you will need to study (2 hour minimum). Use the following rules-of-thumb for budgeting your time:

Easy courses   _____ credit hours x 2 hours = _____ hours

Medium courses   _____ credit hours x 3 hours = _____ hours

Hard courses   _____ credit hours x 4 hours = _____ hours

Total hours per week for school   = _____ hours

Are you currently spending this much time with school? Do you have enough time in your week to accommodate this much time? I would encourage you to make the commitment to set aside this much time for school over the next month. If this does not seem realistic, then you should ask yourself why it is not realistic. It is important to get a handle on your time budget and put into practice what you have learned here. This will set the foundation for your success and your sanity throughout college.

7.3 Budget your time.

Estimate the amount of time each activity will require. Be realistic! You wouldn't just keep writing checks or using the debit card without tracking what has been spent? It's the same with time. Make sure you leave some time slots blank! You must have empty space in your schedule! (You need to eat and sleep!) Here's a fairly typical week for a full-time student working full-time with a family.

Try This! Use the following weekly planner to record your responsibilities outside of school (work, in-class time, study time, family, meals, etc.). Use the "School Rule" to determine study time (on average 3 hours per credit, outside of class).

Time	Monday	Tuesday	Wednesday	Thursday	Friday	Saturday	Sunday
4:00							
5:00							
6:00							
7:00							
8:00							
9:00							
10:00							
11:00							
12:00							
1:00							
2:00							
3:00							
4:00							
5:00							
6:00							
7:00							
8:00							
9:00							
10:00							
11:00							
12:00							
1:00							
2:00							
3:00							

Use the following formula to determine the number of available hours you have for the week.

Free Hours for the Week =
168 – _____ hours (from your schedule) – 56 hours (sleep)

The number of *free* hours for the week represents the total number of hours outside of work and school that you have for decompressing or rejuvenating. You're setting yourself up for failure if you don't allow for quiet time or empty space in your schedule to revitalize your mind and body.

A student came to me mid-way through the semester feeling overwhelmed. I asked them to write down their regular obligations, and I showed them how quickly the hours for the week were disappearing:

Time Budget:
      168 hours / week (total)
      —40 hours / week (work)
      —12 hours / week (class time, if you're taking 12 credits)
      —36 hours / week (study time, if you're taking 12 credits)
      —56 hours / week (sleep)
      —14 hours / week (meals)
      10 hours remaining

This left her with only 10 "free" hours for the entire week. I say *free* in quotes, because this really wasn't free time at all, she didn't include her commute time to work and to school, her hour of walking each week (light exercise), and a number of other activities that nickel-and-dimed her time away. She only had a little more than 1 hour/day to fit everything else into her life! (Notice that had she tried to add another 3 credit course to her schedule, she would've exceeded the number of hours in the week.)

She was wondering why she was so overwhelmed and when we went over her weekly schedule, she realized that certain essential things were suffering (sleep and study time), resulting in lower performance across the board. She realized that certain life-styles changes had to be made.

It is very important to get a realistic sense of your schedule and life-responsibilities. As the gatekeeper of your most valuable resource, time, you must choose wisely what to include and not include in your schedule. Your education is a marathon, not a sprint, to the finish line. Pace yourself so you get it right the first time.

7.4 Learn to say no.

You can apply this rule in nearly any facet of your life. During your time in college, you will find yourself in a number of situations where you will need to say "No I can't go tonight..." or "Sorry, I have to study..." Often times, we say yes because we either don't want to face the drudgery of studying math or we would rather enjoy the immediate satisfaction of doing something fun. The truth of the matter is that by putting off your study time you are actually increasing the amount of time you will need to study. The more time between study sessions the longer it takes to get back what you've lost. Remember to study efficiently, you need to study regularly. It is important to learn to say no nicely to spontaneous invitations that eat up your time, so you can create time for more important things.

7.5 Reward yourself for small and large successes.

Rewarding yourself for achieving both small and large successes can be a great motivator. For each goal you set, include a reward as well for achieving that goal. A goal might be to study math one hour every day for a week. A reward might include giving your self some free time to read, watch T.V., go for a walk, enjoy a favorite comfort food, or hang out with friends. Eventually, you may find that your success is reward enough for maintaining a healthy time management practice.

CHAPTER 8:
DURING CLASS AND NOTE TAKING

8.1 Introduction

"Opportunities are usually disguised as hard work, so most people don't recognize them."

—Ann Landers

It's important to realize from the beginning how important it really is to attend class. Each day in class is an opportunity. Make the commitment to attend and participate in every class. Whether you are in a face-to-face or online classroom, participation and attendance play a significant role in your academic success.

8.2 Location, Location, Location!

The three most important things to remember in real-estate are: location, location, and location! The first important decision you will make once you walk into your classroom will be where to sit. Avoid the back few rows and the far-right and –left columns. You want to be in the front few rows and toward the middle of the class. There are a couple of reasons for this. First, you want to be able to see the board. Sometimes even the most seasoned instructors partially block the view of students on the far right and left sides of the class. You're paying a lot for this seat; don't waste it on the back row!

The seat diagram on the following page shows premier seats, good seats, and cheap seats. If the classroom is full, then obviously everyone can't get the premier seats; however, the earlier you get to class, the better your chances are of getting a good or premier seat.

Online Classes

If you're in an online class, the analogy would be to make the commitment to *actively* participate in class. The "location" in this case is your visibility. You want to be visible to your instructor through your early and frequent participation in assignments and discussions. Make sure you complete the weekly requirements early so you create time for your instructor to give you sufficient feedback on your work. In some cases, your instructor may allow students the opportunity to revisit their work if there's time (no guarantee, but still a possibility).

Seating Diagram

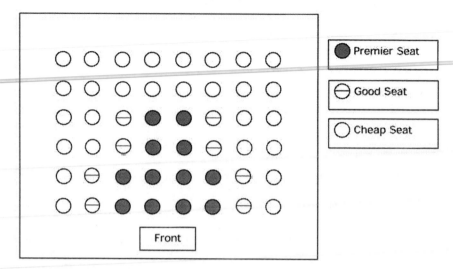

8.3 Three-Ring Binder

- **Front Section** On the first page include the following information:

  o Your Name, Course Number/Title, Instructor Name & Contact Information, Tutor Center Information, Contact Information of at least 2 classmates.

Name: _____

Course Number/Title: _____

Instructor Contact Information

Name: _____
Email: _____
Office Hours: _____

Math & Statistics Tutor Center Information

Log-in Information (Online): _____
Room Location (Face-to-Face): _____

Student Contact Information (Classmates)

Name: _____
Email: _____
Phone: _____

Name: _____
Email: _____
Phone: _____

This title page can be modified to fit your particular class. Notice the title page includes the contact information of your instructor (you'd be surprised how many students don't know the name of their own math instructor!) The remainder of your notebook should include the following five sections:

1. **Class notes & class work** As the title suggests, you will include all of your class notes and work completed in class in this section. You should keep your notes in chronological order. Notes for each day of class should begin on a new sheet of paper. Be sure to include the dates on every sheet in your notes section. It also helps if you can include a note at the top of each page on which quiz, test, or exam this lecture is covering (refer to your syllabus or ask your instructor to verify this). If you miss class, make sure you make a note on a single sheet with the date so you can correctly place the notes in your notebook from one of your classmates. You should rewrite the notes from your classmate to reinforce the note taking process. This slows you down enough to more thoroughly read the notes as you copy them down in your own handwriting.

    Taking notes is a learning process. You need to become familiar with a style of note-taking that helps you write down what is being said in the classroom. The following tips can help you become a better note-taker.

Note Taking Tips
1. **Start each class with a new sheet of paper.** Be sure to include th date at the top of each sheet.
2. **Leave empy space in your notes.** You want to leave space in your ntoes for future comments and questions. It's easier to read notes that are not crammed together.
3. **Write everything down!** If you get confused or feel lost, keep taking notes—you can always go back and ask a question later. It's important to get everything down on paper even if you don't understand it at the moment. Your notes are your best account of what was covered in class. If you miss something, it can be easily lost and forgotten. Leave ample space in your notes where you get confused and then ask your instructor at the end of class to clarify what was said.
4. **Use abbreviations and symbols.** The less you write, the more you will capture. Some common symbols and abbreviations are included at the end of this book. Add some of your own personal favorites.
5. **Pay close attention to your instructor.** You will find that instructors instinctively give clues to what's of particular importance (what you'll likely see later on a test). Make a note of this using a five-star system (1-star is noteworthy, 2-star is important, ... 5-star means you'll see it on your quiz, test, and final exam!) The 5-star method helps you put emphasis when your instructor emphasizes something in class.
6. **Rewrite your class notes after each class.** I encourage my students to rewrite their notes. This helps them go over the material more thoroughly while it is still fresh in their minds. It also helps keep your notes more organized and easier to read. You can use highlights or colored pencils to make it easier to follow our notes.
7. **Circle or highlight key terms** (or terms you do not understand). Create a glossary of these key terms and review them regularly. Math is a language and you need to keep up with new vocabulary terms.

2. **Vocabulary.** In this section of your notebook, you will include definitions of key terms from your notes, text book, and homework. In addition to using the "official" definition in your notes or text, try to define the term *in your own words.* This helps you better understand the concepts behind the math jargon and that's what is most important. Math is a foreign language and it's very important that you improve your vocabulary in this language by reading the vocabulary section often.

3. **Homework.** Your homework is vital to your success in math and so it deserves its own section. Here you will keep all of your homework assignments whether they are graded or not. Always start each homework assignment on a new piece of paper. The following tips can be useful:

- Include the original statement of the problem with each solution.
- Show all your work and explain each step (where appropriate).
- Box-in your final answer (where appropriate).
- Write in pencil so you can more easily change answers if necessary (neatness is important to your instructor, especially if they're looking at your assignments).

This section of your notebook is reserved only for assigned homework problems. You should keep any additional practice problems in another notebook (or folder) to keep your notebook from becoming unmanageable.

4. **Handouts.** Include all handouts given in class. For instance, the first handout is usually the class syllabus. If you find some additional resources (from the publisher of your textbook, Internet, library, etc.) you can include them here. If you miss a class be sure to check with your instructor or a fellow classmate on whether or not handouts were distributed.

5. **Exams & Quizzes.** Here you will place all of your quizzes and exams (3-hole punched) in chronological order. Include corrections to missed questions on a sheet of paper immediately following the quiz or exam. Write down any questions you have about particular solutions on the quiz or exam and include your instructor's answer to your question here as well. A well-organized exams and quizzes section will make a great study tool for later exams and the final.

CHAPTER 9:
TECHNIQUES FOR STUDYING
AND RETENTION

9.1 Getting to Know Your Text Book

"Math is like love: a simple idea but it can get complicated."

—R. Drabek

Why are math and statistics text books so complicated? Why can't they seem to make things easier instead of more confusing? Why do I really need my text book, it isn't helpful! Actually, this is where we should get off the train. As it turns out, your math text is one of your best resources. It is a treasure trove of information:

- All those confusing mathematical terms are defined.
- Examples include complete solutions to problems similar to the homework problems.
- Answers to half of your homework (and maybe even test questions) are included in the back of your text (usually the odd-numbered problems).
- Chapter Review problems are at the end of each chapter to help you figure out what you know and don't know in the chapter.
- Chapter Test Prep problems are at the end of each chapter to help you prepare for quizzes and tests.

Text books are filled with all kinds of important information (vocabulary terms, examples to show you how to solve homework problems, applications, chapter reviews, and practice tests just to name a few). It's important to know how to use your text. The text should be working with you not against you. Reading a math or statistics text is *not* like reading a book for leisure. So often students try to approach a text book like they would a novel for leisure. This approach will frustrate any student. You should read your math or statistics text slowly, jotting down questions or comments as you read. It is also useful to work out the examples in your text to find solution patterns in your exercises at the end of each section. Remember to read your text before each class. While some of the information in your text may be confusing, you are more likely to understand the in-class lectures. You will certainly be in a better position to ask good questions.

It is very important to know all the features of your text book. Your text book will be of no use to you if you can't find what you are looking for. Moreover, if you don't understand how your book is laid out you are less likely to continue reading it before each class. The following is a list of commonly found 'features' in most standard math text books:

- Table of Contents
- Glossary/Vocabulary
- Examples and Exercises
- Index (Back of the book)
- Answers (odd-numbered problems)

9.2 Quick Study After Class.

Does this ever happen to you? You seem to follow your instructor in class. Then several hours later, or several days later, when you get around to revisiting your class notes and attempting your homework you feel like you've forgotten everything. This is because you have.

There are billions of neurons in your brain storing and processing information. Your neurons are clustered together by synapses. The more you learn, the more connections are made between groups of neurons, to form dendrites. When you are first introduced to a new skill, a weak dendrite formation is made. The more you practice a skill the stronger the dendrite connection becomes. The longer you wait to practice a skill the weaker dendrite bond becomes lost. Hence the phrase,

If you don't use it, you lose it!

How many times do you leave a class feeling like you understood what was being taught? You follow the examples reasonably well. You participate in class. Then after a couple of days when you revisit your notes and try to do some of the homework, you feel lost! You're not alone.

Your brain made connections; however, the connections were not reinforced soon enough. Remember in Chapter 8, we discussed the importance of practicing math frequently and as soon as possible (preferably right after class). It's better to set a routine of math practice everyday for 30–60 minutes, then to hold marathon sessions over the weekend, even if you get less time to study this way.

9.3 Practice, Practice, Practice!

Often times, students do well enough on the homework only to find themselves struggling on tests and quizzes. Why is this? They do not use study techniques that reinforce understanding. In other words, they do not practice enough to make connections in their brain strong enough to last beyond short term memory (less than a few minutes). While practicing math regularly is essential to becoming a successful math student, it is not enough. We also need to know *how* to practice. The remainder of this chapter offers specific study tips and techniques. An important first step is getting to know your text book.

9.4 Note cards.

You'll want to pay attention to this section because a good note card system can improve your test scores by an entire letter grade. Start with 1–2 packs of blank 5 x 7" note cards. On the front of each card, write a vocabulary term, key concept, name of a formula, or statement of a problem. Then on the back write a definition, brief explanation, actual formula, or solution to the problem. Doing this will help you organize a massive amount of information into smaller pieces and give you the opportunity to test your knowledge regularly.

Examples of a good note cards are shown below.

Good Note Card: Vocabulary Term

Front   Back

DISTRIBUTIVE PROPERTY

For real numbers $a, b, \& c,$
$$a(b+c) = ab + ac$$

Notice the key term is written on the front of the note card, while the definition is written on the back. Vocabulary note cards are very useful for memorization.

Good Note Card: Formula

Front   Back

Quadratic Formula
(Solution to
$ax^2 + bx + c = 0$)

$$x = \frac{-b \pm \sqrt{b^2 - 4ac}}{2a}$$

Notice the name of the formula is written on the front of the note card, while the actual formula is written on the back. You can read the name of the formula, try to recite the formula, and then turn to the back of the note card to verify your formula.

Good Note Card: Problem Solving

Front   Back

Solve the equation
$x^2 - 3x = 1.$

Write the equation in standard form:
$x^2 - 3x = 1 \Rightarrow x^2 - 3x - 1 = 0.$
Use the quadratic formula, where
$a = 1, b = -3, c = -1$
$$x = \frac{-b \pm \sqrt{b^2 - 4ac}}{2a} = \frac{-(-3) \pm \sqrt{(-3)^2 - 4(1)(-1)}}{2(1)}$$
$$= \frac{3 \pm \sqrt{9+4}}{2} = \frac{3 \pm \sqrt{13}}{2}$$
$$\boxed{x = \frac{3 + \sqrt{13}}{2}} \text{ or } \boxed{x = \frac{3 - \sqrt{13}}{2}}$$

Notice the statement of the problem is written on the front of the note card, while its solution is on the back. You can read the front this type of note card, try the problem, and then flip it around to check your work. Try to resist the temptation to peek before you've really attempted the solution on your own.

Our last example is to show you a bad note card. There are multiple formulas, a graph illustrating the concept, and a hint on how to solve the problem. This is not a good note card because it you are given too much information on the front (there should only be one concept per note card) and there is information on how to solve the problem (which should only be on the back of the note card).

Bad Note Card: Problem Solving

Front   Back

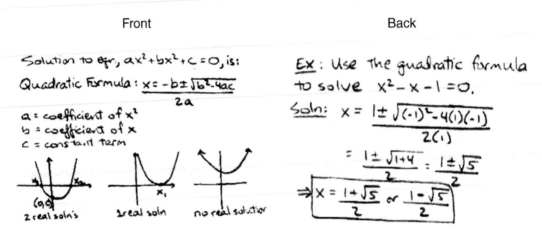

The process of creating the note cards is valuable for a number of reasons. You are reading through the information making determinations about what is important and what is not. Note cards are easy to take with you anytime and anywhere. Students are creative when it comes to studying with note cards. Whether you have 2 minutes or 2 hours to study, note cards can be used in a variety of situations:

- during a meal
- waiting for a ride
- between classes
- during a walk
- in the bathroom
- while you're enjoying a cup of coffee
- before you go to bed

Another great benefit to using note cards is that it reinforces all three primary learning styles discussed in Chapter 4. It supports *visual learning*, since you're looking at the note cards. *Kinesthetic learning* is reinforced in writing the note cards. Reading the note cards out loud or having someone read them to you supports *auditory learning*.

Most students work with a stack of note cards about 2–3 inches thick. A rubber band holds them together and they can be easily carried around even when you're not carrying your note book or text. I would recommend you ask your instructor to briefly review your note cards. If you are worried about taking up too much time, you can show your best 20 cards. By asking for your instructor's feedback, you are not only demonstrating your efforts, but also giving them an opportunity to help you hone your focus. This may even result in some helpful hints for future test items.

Try This! Make 20 note cards for one section in your text book (preferably a section from a current chapter). Include at least one of each type of note card: vocabulary, key concept, formula, and problem. Then explain whether or not you think note cards are a good fit for you.

_____

_____

_____

_____

_____

9.5 Concept Maps

Concept Maps

I have found that students really enjoy making concept maps. It is particularly useful for visual and kinesthetic learners. A concept map is essentially an overview of the important concepts you have covered over a relatively brief period of time (eg, covering a section in your text) in a flowchart. Related concepts and examples are connected by a line. You begin with the section title at the center of concept map. Then write down topics or objectives discussed in that section around the centered section title. It helps if you box in the topics and examples. Figure 1 is an example of a student's concept map for Section 3.2 Introduction to Functions.

Concept maps may seem like a lot of work at first, but there are a number of benefits to creating and using concept maps. Concept maps are great for both visual and kinesthetic learners. This technique improves memory by reinforcing connections between concepts.

9.6 Tutor Center

The longer you wait for help, the harder it gets to ask for it. Most colleges provide free tutoring either face-to-face or through a distance tutoring program. Your instructor should have information about the tutor center. I highly recommend you visit the tutor center (or log into it, if it's a virtual center) and get a feel for the environment. It's a good idea to meet with several different tutors within the first week or two of class to build relationships with the

Figure 1: Example of a Concept Map for a Chapter on Introduction to Functions

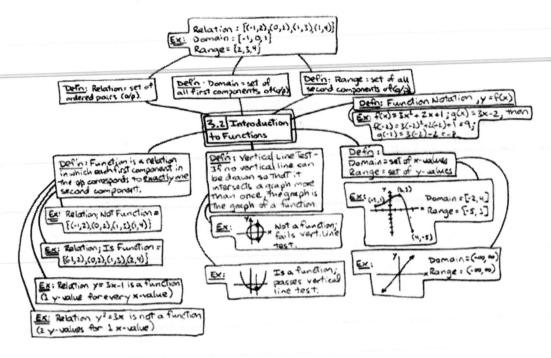

tutors. You will inevitably find a small number of tutors that you prefer and some you do not. Better to find out early than in the middle of finals week!

9.7 Study Groups

One of the most important tips I offer my students is to join a study group. I've found over the years that students who join study groups succeed.

CHAPTER 10:
TEST PREPARATION
AND TEST TAKING

Introduction

Successful test preparation requires a plan. If you don't have a plan, you'll end up wasting hours of your precious time. The first step to creating a study plan is to manage your time (Chapter 7). By starting your test review early, you give your brain time to make connections, resulting in a better understanding of concepts. Getting involved in a study group often gives you different insights and perspectives. If there's one thing I want you to get from book, it's this:

To succeed, you need a plan.

10.1 Make a study plan

Planning ahead will save valuable time. Chose a location where you will have uninterrupted time to study. Be realistic about how often and how frequently you will be able to study. Create a weekly schedule that includes specific times and locations for study sessions. Try to plan frequent 1-2 hour study sessions and avoid single marathon sessions on the weekends or late at night.

Try This! Describe how you currently prepare for tests. Include how often you study and how long your study sessions typically run. Include any specific strategies you currently use. Which are effective and which do you think might stand improvement?

_____

_____

_____

_____

_____

A common misconception among students is that by doing all the homework they should have all the necessary skills to pass the test. Is this realistic? Think for a moment about what you have available during homework and study sessions:

- Pencils
- Scrap paper
- Textbook
- Notes
- Homework assignments
- Calculator
- Classmates, friends, and family
- Tutoring
- Answers in the back of the book
- Complete solutions in your student solutions manual
- Unlimited time

What you have available during the test:

- You
- Pencils
- Scrap paper (maybe)
- Calculator (maybe)

Homework is an important part of succeeding in a math. However, doing your homework and getting ready for a test are two entirely different things. Doing your homework is just one part of getting ready for your test. If you do not stick to a *complete study plan*, your test performance will suffer.

Bring all of your resources to your study session. This means having access to the tutor center, while you study. If there is a tutor center on campus, try to hold your study session in or near the center. If you are using a distance tutoring program, then bring all the information you will need to access your tutor to your study session.

Don't forget your study tools (Flash Cards, Concept Maps, and Cheat Sheets). Study with a classmate or in a small group. Using your study tools with another person reinforces your auditory learning and improves retention.

10.2 The Art of Test Preparation

Taking a test is like performing in a symphony concert in front of a live audience! Doing your homework, reading your text, and reviewing your notes is like reading the manuscript. You need the manuscript to learn the music. However, most musicians will admit that just reading the manuscript will not result in a good performance. In fact, if a musician relies too heavily on simply reading the music they won't remember the music at all. As an important step in finalizing their preparation, symphonies often perform a dress rehearsal in front of a live audience. Mistakes are often made during dress rehearsals and the musicians have time to work on improving their performance. Many students miss out on their own *dress rehearsal* during test preparation. They study hard and do all the homework problems, but don't realize how much they've come to rely on the textbook and notes during their preparation.

10.3 Practice Tests – Dress Rehearsal

Most text books have practice tests at the end of each chapter. However, your instructor may give exams that do not resemble the chapter tests in your text. If this is the case, then you will use your class notes, home work, and text book to help create your own practice tests. For each practice test, choose a sample of 20 questions (or ask your instructor how many questions will be on the quiz or exam to determine how many you should include on your practice test). Be sure to include a sample of each important topic that will be included on the actual test. For your first practice test, you may want to choose only odd-numbered problems from your text so you can use the answers in the back of the book. As you are writing the problems for your practice test, reference where the problem came from in your text so you can go back and check your work or review a concept in your text when you're done with the practice test.

Making and taking practice tests or quizzes is a great studying technique in general. I ask my students to make practice quizzes (10 questions each) after they have completed each section in their text. Making a practice test encourages you to go over important concepts closely. By taking practice tests, you make the entire process of test taking more familiar and therefore a little less intimidating. Making and taking practice tests and quizzes on a regular basis builds confidence and reduces math anxiety.

10.4 More Thoughts on Test Preparation

Test preparation begins on the first day of class. You must make the commitment to do your homework and complete it on time. Use your homework along with your class notes and text to create your flash cards, vocabulary, and concept maps.

An important tip for successful test preparation is knowing what to study. Two weeks prior to the test date, ask your instructor to describe your test or quiz. Many instructors don't mind discussing the test, but it's important to know when to ask and what to ask. By asking these questions a couple of weeks prior to the test, you are showing commitment and dedication to your math course. Here are some examples of good questions to ask:

- Which course materials will the test cover? (Class notes, text, or a combination of both).
- Is the test cumulative? Which sections or chapters will be emphasized on the test?
- What kinds of questions will be on the test? (multiple choice, true/false, open response, etc...)
- How many questions will be on the test?
- How much time will we have to take the test?
- Will a calculator be allowed?

Show your instructor your flash cards and concept maps. Ask your instructor for feedback on your concept map. Listen closely, the suggestions might show up on your test. Identify homework questions that were particularly challenging and ask your instructor if they are relevant for the test. Your instructor will appreciate that you're planning ahead for the test.

One week prior to the test, increase the amount of time you spend reviewing your homework and notes from 20 minutes to at least 30 minutes a day (preferably 1 hour a day). Use your flash cards as often as possible. Review your notes and vocabulary terms. Begin writing your practice test.

A few days before the big day, take your practice test. By taking it a few days before the actual test day you give yourself enough time to review any concepts you missed. This also gives you some time to regroup if you did poorly on your practice test. On the day before the test, you should be merely brushing up on a few details and giving yourself plenty of sleep. The day of the big test, eat a nutritious breakfast and remember to bring your hard candy!

10.5 Test Day – The Big Performance

- **Arrive Early** – Try to arrive at the location of your test at least 10 minutes prior to the test so you can get comfortable with your surroundings. Bring a light snack and beverage if allowed for energy and comfort.

- **Relax** – (Visualization – my cat relaxing on the bed warm sun coming through the window. A summer vacation... Whatever works for you...) When the test is distributed, take a deep breath and relax. Read the directions carefully. Browse the entire quiz or test quickly before beginning work.

- **Read the directions carefully** – You want to make sure you're answering the question that is being asked.

- **Pace yourself** – If there are 10 questions on the test and you have 60 minutes, then on average you want to spend at most 6 minutes on each question. Some questions will require less time and others will require more, but overall you want to budget your time wisely.

- **Check your work** – Time permitted; once you've completed the test, check your work and your answers. Do not turn in your test early. If you have time remaining, double-check your work. You do not get extra credit for finishing early, but you might catch a careless mistake!

Conclusion

"The future depends on what we do in the present."

—Gandhi

I believe Gandhi is trying to tell us that what we do now affects our future. That is, we get out of life what we put into it. By taking our education seriously, managing our time, being disciplined, and maintaining a positive attitude we are creating an environment that will *receive* our success. This enables us to *achieve* our success.

This book is about more than just study skills it's about *success* skills. It is about empowering you to take full responsibility for your success. The general principles in this book can be used in many areas of your life. Take from it what works for you. Focus on what brings you success. And be open the success that will come.

Math Short-Hand for Faster Note Taking

Short-hand	What it means
add.	the word "addition"
b/c	because
CLT	combine like terms
Def:	definition
div.	the word "division"
dst pr	distributive property
dist.	Distance
Eqn:	equation
Ex:	example
exp	exponential or exponent
fn	function
frac	fraction
ineq	inequality
log	logarithm
lin	linear
mult.	the word "multiplication"
p.	referencing a page number in your textbook
pf	proof
parens	parentheses
proba	probability
soln	solution
sqrt	square root
sub.	the word "subtraction"
thm	theorem
w/	with
w/out	without
wolog	without loss of generality
?	use this to mark things you don't understand
!	use this to mark things that are important
!!	use this to mark things that will be on the test
=>	implies

References

Birenbaum & Nasser, 2006
http://www.sciencedirect.com/science?_ob=ArticleURL&_udi=B6VFW-4JFF047-6&_user=961261&_rdoc=1&_fmt=&_orig=search&_sort=d&view=c&_acct=C000049394&_version=1&_urlVersion=0&_userid=961261&md5=4da83a0146864a22282697ed8e6abb0d#sec4.2